YOUR KNOWLEDGE HAS VALUE

- We will publish your bachelor's and master's thesis, essays and papers

- Your own eBook and book - sold worldwide in all relevant shops

- Earn money with each sale

Upload your text at www.GRIN.com and publish for free

Bibliographic information published by the German National Library:

The German National Library lists this publication in the National Bibliography;
detailed bibliographic data are available on the Internet at http://dnb.dnb.de .

Imprint:

Copyright © 2016 GRIN Verlag, Open Publishing GmbH
Print and binding: Books on Demand GmbH, Norderstedt Germany
ISBN: 9783656989769

This book at GRIN:

http://www.grin.com/en/e-book/323712/spatial-variation-of-reference-evapotran-
spiration-and-its-influence-on

L.R. Singo, P. M. Kundu, J.O. Odiyo, F.I. Mathivha

Spatial Variation of Reference Evapotranspiration and its Influence on the Hydrology of Luvuvhu River Catchment, Limpopo Province, South Africa

GRIN Publishing

Table of Contents

SPATIAL VARIATION OF REFERENCE EVAPOTRANSPIRATION AND ITS INFLUENCE ON THE HYDROLOGY OF LUVUVHU RIVER CATCHMENT, LIMPOPO PROVINCE, SOUTH AFRICA

L.R. SINGO[1], P.M. KUNDU[1], J.O. ODIYO[1] & F.I. MATHIVHA[1]

ABSTRACT

Evapotranspiration (ET) is regarded as the largest mode of water loss in arid and semi-arid areas and is critical for accurate predictions of water exchange and crop productivity. Analysis of spatial and temporal fluctuations of evapotranspiration is therefore crucial to understand the coupled water and energy cycles in arid and semi-arid environments. This study estimated reference evapotranspiration (ET_o) to study its influence on the hydrology of Luvuvhu River Catchment using a physically based model. The ET_o plays a key role in irrigation systems design, water management under irrigated and rainfed production. The Penman-Monteith equation which is widely used in water resource management and planning was used to estimate ET_o. Simulation of ET_o was performed using CROPWAT 8.0 software. This algorithm is useful in simulating water resource management scenarios at different spatial and temporal scales under a wide range of environmental conditions. To understand the impact of climatic characteristics in the formulation of evapotranspiration model, seasonal variation of different meteorological parameters such as wind speed, solar radiation, temperature and humidity was analysed. Results showed the spatial and temporal distribution of ET_o for different climatic stations in the study area with peaks in summer months. Minimum values of ET_o were observed during the dry months. Results from the simulations showed that the areas with higher ET_o values were near rivers and streams, which generally have more abundant vegetation. Areas with low ET_o values were relatively dry, where pasture and grasslands dominated the landscape. Correlation results showed that no relationship exists between stream flow and ET_o ($r = 0.36$) in the study area, hence, a significant relationship exists between rainfall and ETo ($r = 0.86$). The study recommends the use of CROPWAT model for computing ET_o under arid and semi-arid climatic conditions for water resource management and planning.

Keywords: CROPWAT, catchment, Penman-Monteith, reference evapotranspiration, simulation.

1   INTRODUCTION

Evapotranspiration (ET) is the sum of evaporation and plant transpiration from the Earth's land surface to atmosphere. The ET is influenced by local conditions that range from precipitation and meteorology to soil moisture, plant water requirements and the physical nature of the land cover (Abdullahi [1]). Hence, ET is a key process within the hydrological cycle and possibly the most difficult component to determine, especially in arid and semi-arid areas where a large proportion of low and sporadic precipitation is returned to the atmosphere via evapotranspiration (Jovanovic *et al.* [2]). In these areas, ET is roughly equal in magnitude to precipitation on timescales longer than seasons (Reynolds *et al.* [3]); and is regarded as the largest mode of water loss critical for accurate predictions of water exchange and crop productivity (Moiwo *et al.* [4]). Knowledge of ET is therefore critical for sustainable water resources management and balanced water supply among industrial, domestic, ecological, and agricultural sectors (Moiwo *et al.* [5]). Whereas climate and hydrological variables such as precipitation and runoff are measured with reasonable accuracy, ET is difficult to quantify especially at the basin/regional scale. Estimates of ET are therefore required in the design of reservoirs, irrigation systems, scheduling and frequency of irrigation and water balance and simulation studies (Moiwo *et al.* [5]).

In cultivated semi-arid regions such as the Luvuvhu River Catchment (LRC) where rainfall occurs on limited basis, estimation of ET is crucial for management of water resources. As a result of spatial variability of landscape characteristics such as topography, soils, land use/cover, and vegetation, ET will likely exhibit a spatial variation in semi-arid catchments (Güntner & Bronstert [6]). In these areas, measurements of ET are rarely available. Evaporation pan coefficients from Symon's pan and Class-A pan have been used by the Department Water Affairs and Sanitation (DWAS) and South African Weather Services (SAWS), respectively, to estimate total annual evaporation losses in the study area. These methods are often not reliable for estimating the seasonal variation in evaporation losses due to heat storage in deep lakes. The DWAS [7] estimated that the highest evaporation (about 60%) occurs during rainfall season while lowest evaporation (about 40%) occurs in dry season. The high evaporation significantly reduces effective rainfall, runoff, soil infiltration, groundwater recharge and causes water loss from water bodies leading to an increase in the concentration of sediments.

In the present study, attempts were made to estimate reference evapotranspiration (ET_o) and study its influence on the hydrology of LRC using a physically based model. The ET_o values can be considered equal to evaporation from a large body of water, such as a pond or lake. However, for smaller, shallower bodies of water this relationship does not apply. Thus, the ET_o is the evapotranspiration rate from a reference surface, not short of water; and is a hypothetical surface with extensive green grass cover with specific characteristics (FAO [8]). It expresses the evaporating power of the atmosphere at a specific location and time of the year and does not consider the crop characteristics and soil factors. The ET_o plays a key role in irrigation systems design, water management under irrigated and rainfed production (FAO [8]). The ET_o was used to study the evaporative demand of the atmosphere independently of crop type, crop development and management practices. The site specific water requirement can be estimated, if climatic data recorded at weather station are available in different sites of a region. The *ET_o* component takes a time series input of meteorological data that include radiation, temperature, vapour pressure deficit, and the rate of increase of saturation vapour pressure with temperature.

With improvements in computer technology and simulation techniques, mathematical models are now widely used to predict the variations in ET in the absence of measurements, given meteorological data describing variations in the climate, and land use data describing variations in the vegetation cover. The ET estimate is now frequently obtained using simulation models the Food and Agricultural Organization (FAO) based Penman Monteith type estimates (Mcglinchey & Inman-Bamber [9]). Satellite sensors are also being used in catchment hydrology to estimate and describe the dynamics of ET. A recent study by Jovanovic *et al.*, [2]) used satellite-derived data to estimate and describe the dynamics of ET in South Africa. In the LRC, estimation of ET with computer technology and simulation techniques has been rare. However, the study's ability to use a physically based model to estimate ET offered the opportunity to understand how ET varies across space and time in the catchment. Time series analysis was also used to estimate variation of ET, surface runoff and stream flow. The only process-based model that is widely used, and that accounts for the influence of vegetation on the ET regime, is the Penman-Monteith formula (Monteith [10]), which is widely used to compute ET_o.

Allen *et al.* [11] proposed the CROPWAT 8.0 to calculate ET_o by the Penman-Monteith method. CROPWAT is a computer model for the calculation of crop water requirements and irrigation requirements from existing or new climatic and crop data. It has an option to calculate ET_o from these data. The Penman-Monteith equation assumes the reference crop evapotranspiration as that from a hypothetical crop with an assumed height of 0.12m, having a surface resistance of 70 s/m and an albedo of 0.23, closely resembling the surface of green grass of uniform height, actively growing and adequately watered (Allen *et al.* [11]). In the present study evapotranspiration was determined as daily values that represented the removal of water from the soil by evaporation and plant transpiration using CROPWAT 8.0 software developed by the Food and Agriculture Organization (FAO).

2   THE STUDY AREA

The LRC shown in Figure 1 is located in Vhembe District of the Limpopo Province in northeastern South Africa, between latitudes 22°17'34"S and 23°17'57"S and longitudes 29°49'46"E and 31°23'32"E, and covers an area of about 5941 km^2 (DWAS [7]). The catchment originates from Soutpansberg Mountains and drains into the Limpopo River, which extends into Mozambique, and is generally semi-arid and becomes arid as it progresses eastward. The climate of the area is largely influenced by the Intertropical Convergence Zone (ITCZ), modified by local orographic effects. The region has two seasons; a rainy season from October to April, and a dry season (long) from May to September. Rainfall distribution in the catchment is classified as unimodal, having a rainy season predominantly between the months of October to January with the average annual rainfall of about 200-400 mm (DWAS [7]). The mean annual temperature ranges from 18°C to 40°C with high variability. Maximum temperatures are experienced in January and minimum temperatures occur on average in winter. Average wind speed in the catchment is approximately 11km/h with high-speed winds occurring occasionally with long intervals.

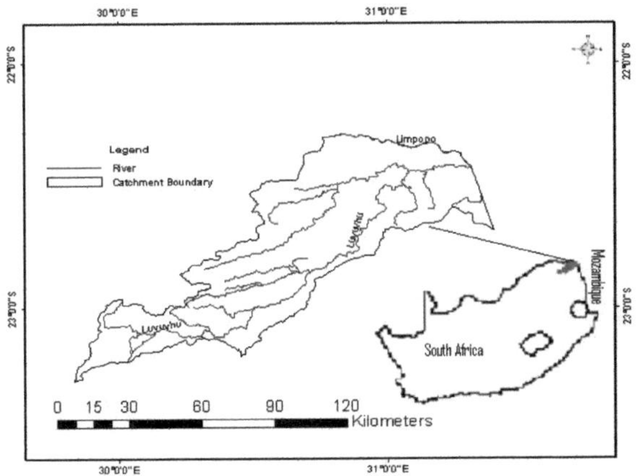

Figure 1: The study area

Topographically, the catchment consists of a relatively rolling landscape, which gives rise to shallow storage dams, which have large water surfaces exposed to evaporation. The topographic feature that characterizes the study area is the Soutpansberg mountain range in the east of the catchment, which reaches an elevation of 1,700m above mean sea level before dropping off into the Limpopo River Valley. The general terrain of mean ridge height approximately 800-1200m is common in some places while in some places the peak reached above 1500m. The predominant soils in upland areas are Leptosols while the lowlands are dominated by Vertisol and Acricsols (FAO [8]). Land use practises largely vary from plantation forests interspersed with large scale macadamia and banana plantations in the headwaters interspersed with small farm holdings for subsistence agriculture. Land cover degradation in the catchment is a major cause of environmental concern. The headwaters are gradually getting depleted due to uncontrolled anthropogenic activities by the rising human population (DWAS [7]).

3 MATERIAL AND METHODS

3.1 Hydrometeorological Data

The daily hydrological and meteorological data (1960-2014) used were all obtained from public sources-the Department of Water Affairs and Sanitation (DWAS), South African Weather Services (SAWS), and the Agricultural Research Council (ARC). The data include daily data on rainfall from eight rain gauges; stream flow and surface runoff from eight flow gauging stations; and ET_o climate data obtained from six weather stations as shown in Figures 2 to 4. ET_o climate data include minimum and maximum temperature, wind speed, relative humidity, sunshine hours and solar radiation. The data were subjected to quality check for missing data, consistency and stationarity.

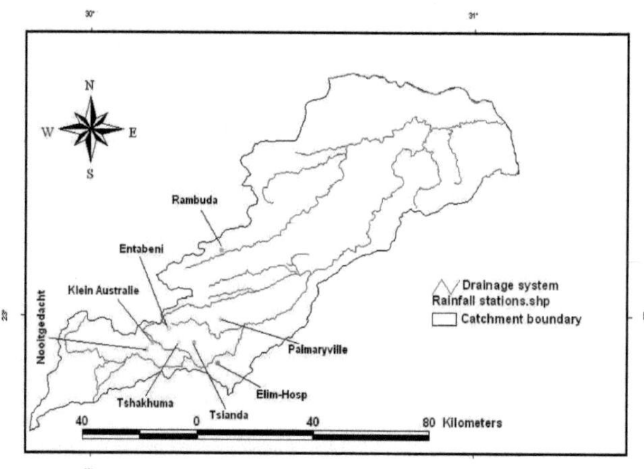

Figure 2: Rain gauges

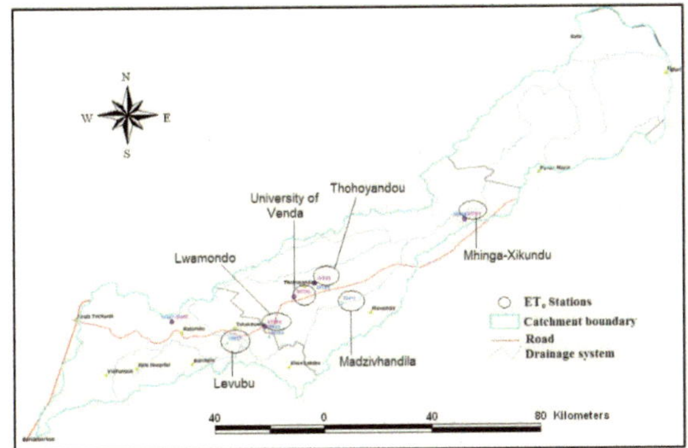

Figure 3: Weather stations for ET_o analysis

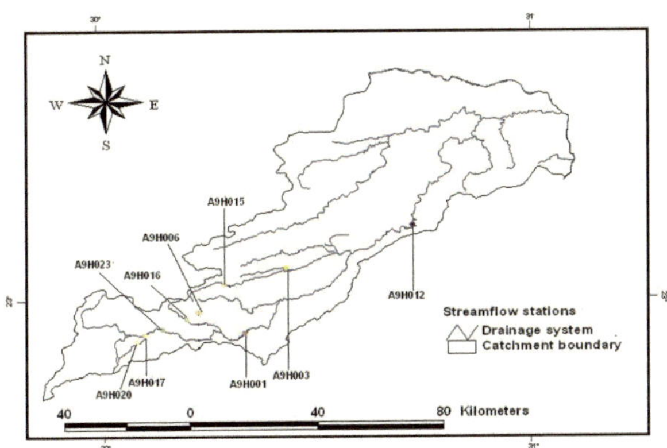

Figure 4: Flow gauging stations

3.2 Evapotranspiration Analysis

Daily ET_o data were calculated by the Penman-Monteith method using full set of climatic data, consisting of minimum temperature, maximum temperature, wind speed, relative humidity, solar radiation, and sunshine hours of daily records of the observatory as input to CROPWAT 8.0. Other data that was needed to compute radiation and sunshine hours for use in ET_o include soil heat flux, vapour pressure, saturated vapour pressure, and vapour pressure deficit. The CROPWAT 8.0 calculates the ET_o by Penman-Monteith eqn (1) as:

$$ET_o = \frac{0.408\Delta(R_n - G) + \gamma\frac{900}{T+273}u_2(e_s - e_a)}{\Delta + \gamma(1 + 0.34u_2)} \tag{1}$$

where ET_o is the reference evapotranspiration (mm/day); R_n is the net radiation of the crop surface (MJ/m^2/day); G is the soil heat flux density (MJ/m^2/day); T is the mean daily air temperature at 2 m height (^{0}C); u_2 is the wind speed at 2 m height (m/s or km/day); e_s is the saturation vapour pressure (kPa); e_a is the actual vapour pressure (kPa); (e_s-e_a) is the saturation vapour pressure deficit (kPa); Δ is the slope vapor curve (kPa/^{0}C); and γ is the psychtometric constant (kPa/^{0}C). However, data on these variables cannot be used as input in Penman-Monteith climate data table in CROPWAT 8.0 model; rather they are used to derive parameters such as radiation, humidity and sunshine hours in stations with limited data.

3.3 Statistical Analysis

The temporal variation in ET_o and hydrological variables were assessed using time series analysis based on linear regression functions. The spatial variation between ET_o and hydrological variables was assessed using the coefficient of determination (r^2) of regression functions and linear correlation coefficient (r). All these statistical indicators were used to compare the relationship and variation between ET_o and the observed hydrological variables, and to test for statistical significance. A linear regression provided the formula for estimating such relationship as:

$$Y = ax+b \tag{2}$$

where x is the explanatory variable and y is the dependent variable. The slope of the line is b, and a is the intercept.

$a = (\Sigma Y - b (\Sigma X)) / N$
$b = (N\Sigma XY - (\Sigma X)(\Sigma Y) / (N\Sigma X^2 - (\Sigma X)^2)$
where, N = number of values or elements; ΣXY = sum of the product of first and second scores; ΣX = sum of first scores; ΣY = sum of second scores; and ΣX^2 = sum of first scores

The coefficient of determination (r^2) was computed as the value of squared correlation coefficient that gives the proportion of the variance (fluctuation) of one variable that is predictable from the other variable. A linear correlation coefficient (*r*) used to measure the strength of the relationship between has the following eqn (3) as provided by Gordon *et al.* [12]:

$$r_{xy} = \frac{\sum\left(x_i - \bar{x}\right)\left(y_i - \bar{y}\right)}{\sqrt{\sum\left(x_i - \bar{x}\right)^2\left(y_i - \bar{y}\right)^2}} \tag{3}$$

where r_{xy} = correlation co-efficient of two variables

x_i, y_i = sum of individuals

$\bar{x}, \bar{y}$ = historical means

4   RESULTS

4.1 Hydrometeorological Analysis

Rainfall data for the study area was used to provide an indication of the variation in precipitation measured by all of the selected rain gauges across the catchment. Past studies have shown that rainfall intensity can have a strong influence on ET. A study by Kurc & Small [13] has shown that ET often consumes a large part of precipitation and the amount and timing of ET can strongly affect stream flow and groundwater recharge in arid and semi-arid climates. The variation in monthly average rainfall measured at each station is illustrated in Figure 5. The figure shows that the study area experienced a unimodal rainfall season that extends from October to March with peaks either in January or February. Rain gauge at Entabeni-Bos had the highest rainfall values in the study area ranging from 32.8 mm during the dry season to 354.7 mm during the rainy season. The station is located along the Entabeni Forest, an Afro-temperate Mist-belt Forest that makes for an interesting combination of Afromontane and tropical vegetation elements.

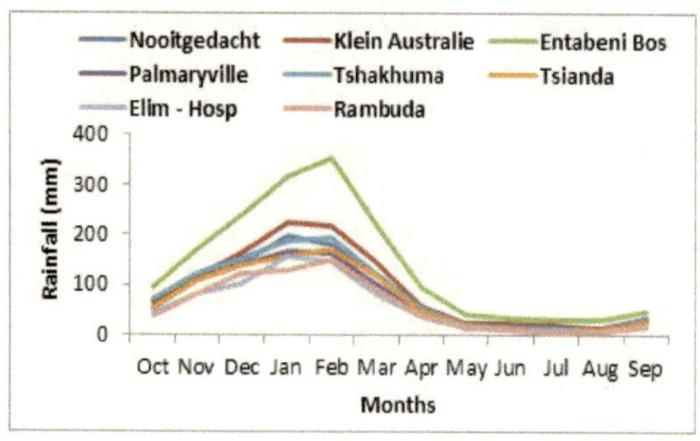

Figure 5: Average monthly rainfall for the study area

Stream flow patterns in the catchment showed spatial and temporal variations as shown in Figure 6. Both gauging stations revealed unimodal peaks centred on February and March while decreasing trends with lesser amounts were experienced during the dry season from May to September. Station A9H001, which has a history of being seriously impacted by the construction of Albisini Dam during the early 1950s and subsequent irrigation abstractions, had relatively high stream flows. Davies & Day [14] estimated that South African rivers experience high water recharge during the summer rainfall season, resulting in abnormal stream flows. However, much of the summer rainfall recharges the groundwater reservoir and is not lost to the system but feeds the rivers during the winter period of water deficit.

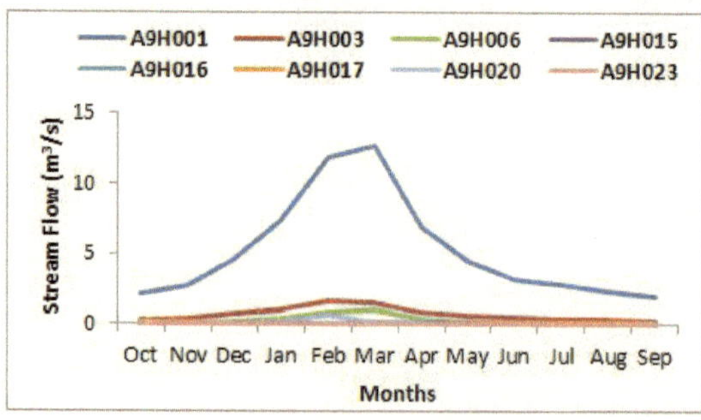

Figure 6: Average monthly stream flow for the study area

For each of the meteorological records used, standard equations were used to derive daily values for solar radiation and sunshine hours following Allen *et al.* [11]. These values were then used in the Penman-Monteith equation to compute ET_o. Figure 7 and 8 shows the computed solar radiation and sunshine hours data used for ET_o analysis. The data showed greater variability during the rainy season from October to March while lesser amounts were experienced in dry months. The annual averages of sunshine hours ranged from 1.6 hours/day to 6.8 hours/day whereas solar radiation ranged from 11.8 $MJ/m^2/d$ to 22.9 $MJ/m^2/d$. It should be noted that not all available energy is used to vaporize water as part of the solar energy is used to heat up the atmosphere and the soil profile.

Values for relative humidity, wind speed, as well as minimum and maximum temperatures obtained from meteorological stations were also made available and used as input in the Penman-Monteith method. The study area experienced maximum wind velocities of 129 km/day (1.5 m/s) and a minimum of 80 km/day (0.96 m/s). High winds tend to increase the rate of evapotranspiration as the rate of evaporation increases with the moving air. Such wind will also clear the air of any humidity produced by the plants' transpiration thereby increasing the rate of evapotranspiration. During the rainy season, average relative humidity was above 60%. Evaporation and transpiration rates tended to drop to low levels when the air around the plants was too humid. Average minimum temperature ranged from 7.5 °C to 20.3 °C with maximum range of 23.5 °C to 32.6 °C. There appears to be a relationship between the increase in temperatures during the rainy reason and the increase in evapotranspiration rates in the study area. This could be due to the fact that high amount of energy was available during the summer months when temperatures were high.

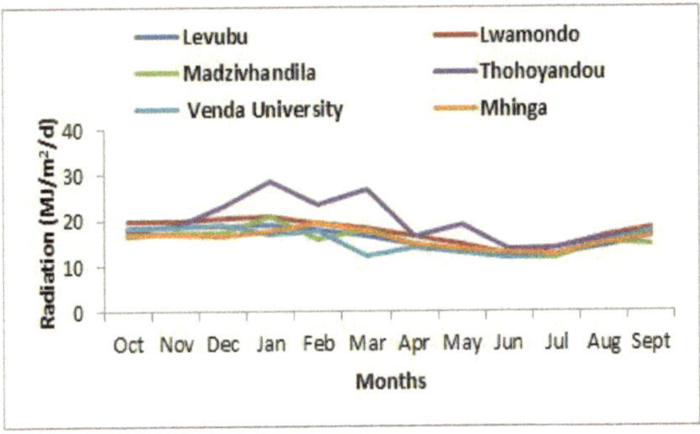

Figure 7: Average monthly solar radiation

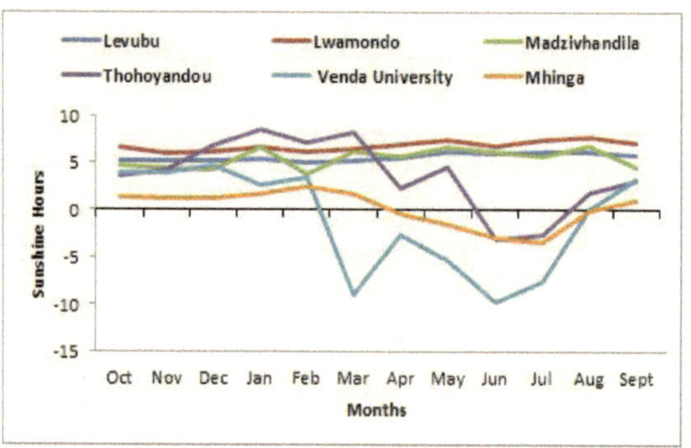

Figure 8: Average monthly sunshine hours

4.2 Reference Evapotranspiration Analysis

Figure 9 shows the distribution of ET_o for the study area. The onset of the rainy season in October at each climate station starts with relatively high ET_o values which then increases gradually until January. Thereafter, ET_o values declined gradually from February and reached minimum in June. It then rose again from July towards the early rainfall season in October. The peak ET_o months were recorded in summer months of January, February and March, throughout for different climatic stations. During the rainy season, Levubu, Lwamondo, Thohoyandou, and Venda University (Univen) observed the high ET_o values of more than 4.0 mm/day. These areas are situated at the middle section of the catchment surrounded by many rivers and dams. The lowest ET_o values were observed during the dry months at Mhinga and Madzivhandila as shown in Figure 5. Mhinga and Madzivhandila areas are located at the lower section of the catchment and are surrounded by natural veld for grazing camps.

Schaffer [15] showed that high evapotranspiration occurs when high available energy interacts with high soil moisture and robust plant health. Areas with the lowest ET_o rates were relatively dry areas, where pasture and grasslands dominate the landscape. Conversely, areas with higher ET_o rates were near rivers and streams, which generally have more abundant vegetation. Mölders [16] showed that evapotranspiration was low in deforested areas due to reduced surface roughness length and it significantly decreased net radiation. Such low ET_o subsequently modifies the water and energy cycles within the deforested area. In areas that are characterized by low rainfall, the rate of evapotranspiration is low as some water will be lost due to percolation and surface runoff.

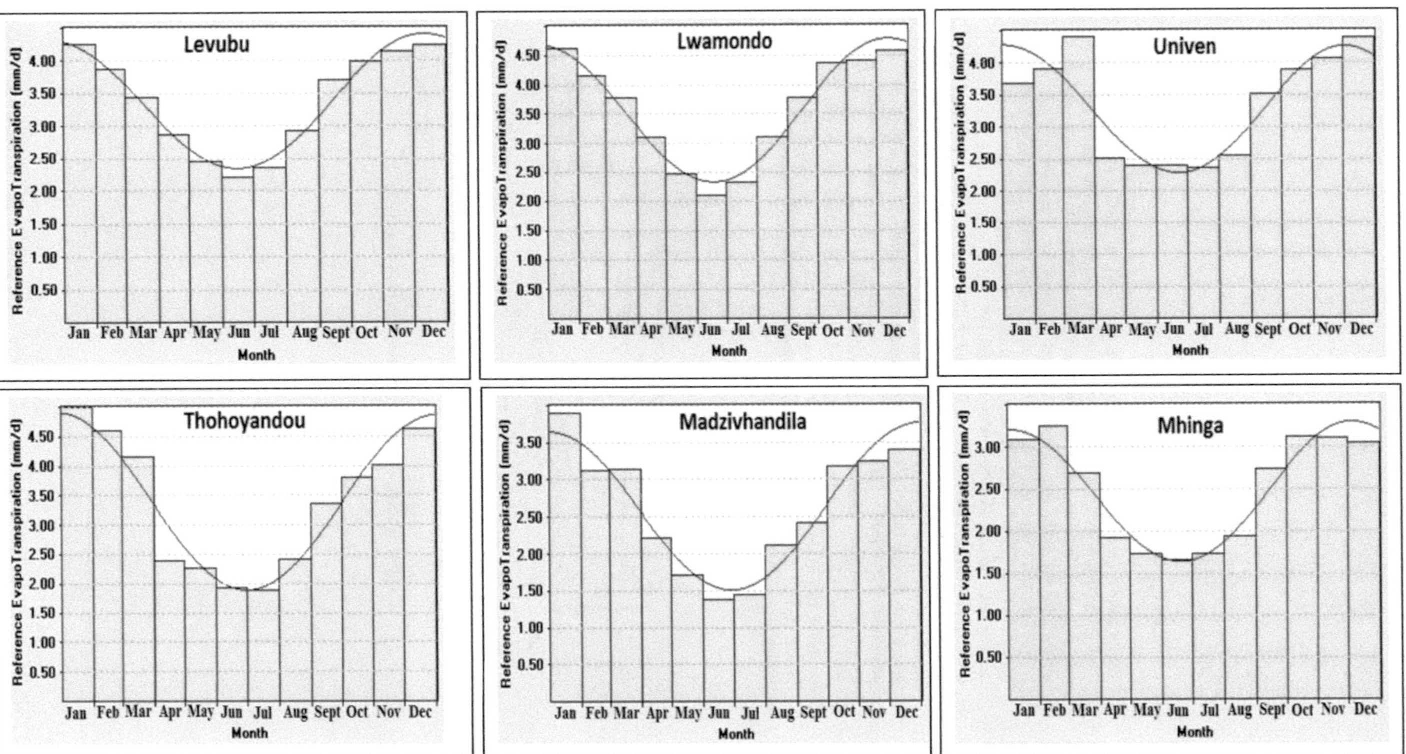

Figure 9: Average monthly ET_o for the study area

4.3 Statistical Analysis

The influence of evapotranspiration on the hydrology of LRC was assessed through regression analysis between ET_o, rainfall, and stream flow data. For each parameter, the data for all stations were combined to produce a single data representing the study area. Figures 10 and 11 show the regression analysis between the three parameters. Rainfall and stream flow reveal decreasing trend during the study period while ET_o shows horizontal trends. Results of correlation analysis detected a weak relationship between ET_o and stream flow ($r = 0.36$), and a significant correlation between ET_o and rainfall ($r = 0.86$). This show that no relationship exists between stream flow and ET_o in the study area, hence, a significant relationship exists between rainfall and ET_o. Rainfall and ET_o are regarded as the two hydrological variables reflected by climate change, water use efficiency, and effective irrigation. The coefficient of determination which explained how much of the variability revealed r^2 values of rainfall, ET_o, and stream flow; each having variances of 0.1074, 0.0071, and 0.5543 respectively, as shown in Figures 10 and 11. It can be assumed that a 10,74% decrease in annual rainfall may have led to a 55.4% decrease in stream flow, but only a 0,71% increase in ET_o.

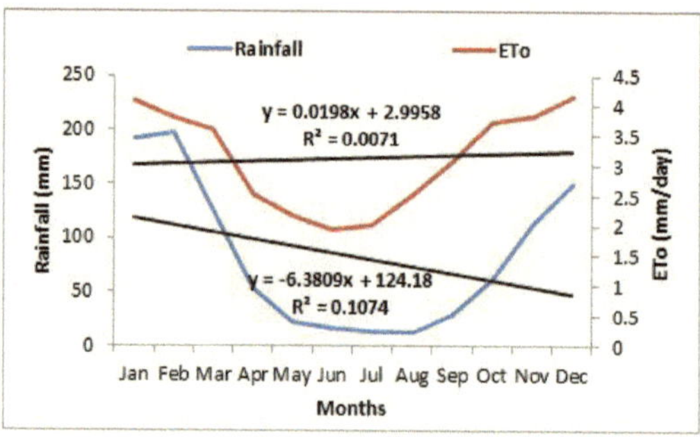

Figure 10: Regression analysis between rainfall and ET_o

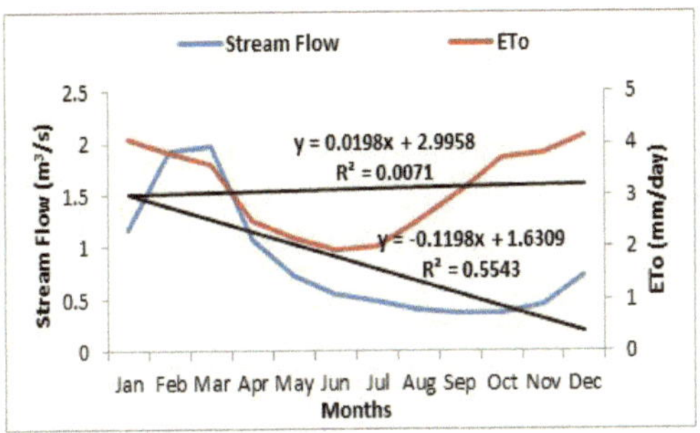

Figure 11: Regression analysis between stream flow and ET_o

5 CONCLUSIONS

The influence of ET_o on the hydrology of LRC was quantified using a number of statistical methods. The variations in the ET_o were estimated through the Penman-Monteith method while trends between ET_o and hydrological variables were investigated using regression analysis. The characteristics of ET_o in the catchment were addressed, and its relationship between two hydrological variables, rainfall and stream flow, was explained. Results showed spatial variation of ET_o in the catchment. Levubu, Lwamondo, Thohoyandou, and Venda University (Univen) observed the high ET_o values of more than 4.0 mm/day while the lowest ET_o values were observed during the dry months at Mhinga and Madzivhandila areas. The estimates of ET_o can therefore be used in the planning process for irrigation schemes to be developed in the catchment as well as to manage water distribution in existing schemes. Results of correlation analysis detected a weak relationship between ET_o and stream flow ($r = 0.36$), and a significant correlation between ET_o and rainfall (r = 0.86). Results showed that a 10,74% decrease in annual rainfall may have led to a 55.4% decrease in stream flow, but only a 0,71% increase in ET_o.

References

[1] Abdullahi, A.S., Ahmad, D., Amin, M.S.M. & Aimrun, W., Spatial and Temporal Aspects of Evapotranspiration in Tanjung Karang Paddy Field, Peninsular Malaysia. *International Journal of Science, Engineering and Technology Research* 2(2): 473-479, 2013.
[2] Jovanovic, N., Mu, Q., Bugan, R.D.H. & Maosheng, Z., Dynamics of MODIS evapotranspiration in South Africa. *Water SA* 41 (1): 79-90, 2015.

[3] Reynolds, J., Kemp, P. & Tenhunen, J., Effects of long-term rainfall variability on evapotranspiration and soil water distribution in the Chihuahuan desert: A modeling analysis, *Plant Ecol.* 150: 145-159, 2000.

[4] Moiwo, J.P., Yang, Y., Xiao, D., Yan, N & Wu, B., Comparison of ET Watch-estimated evapotranspiration with that from combined GRACE and precipitation measurement data in Hai River Basin, North China. *Hydrological Sciences Journal* 56(2), 249–267, 2010.

[5] Moiwo, J.P., Tao, F & Lu, W., Estimating soil moisture storage change using quasi-terrestrial water balance method. *Agricultural Water Management* 102(1): 25–34, 2011.

[6] Güntner, A. & Bronstert, A., Representation of landscape variability and lateral redistribution processes for large-scale hydrological modeling in semi-arid areas, *Journal of Hydrology*, 297: 136-161, 2004.

[7] DWAS, Luvuvhu and Letaba Water Management Area: *Overview of water resources availability and utilisation.* Department of Water Affairs and Forestry, Government of South Africa, 2003.

[8] FAO, *Drought impact mitigation and prevention in the Limpopo River Basin: A situation analysis.* A report prepared by the FAO Subregional Office for Southern and East Africa Harare, FAO, Rome, Italy, 2004.

[9] McGlinchey, M.G. & Inman-Bamber, N.G., Robust estimates of evapotranspiration for sugarcane. *Proceedings of the Annual Congress - South African Sugar Technologists' Association*: p245-249, 2002.

[10] Monteith, J.L., Evaporation and environment. *Symposia of the Society for Experimental Biology*, 19: 205-224, 1965.

[11] Allen, R.G., Pereira, L.S., Raes, D. & Smith, M., *Crop evapotranspiration: guidelines for computing crop water requirement.* FAO Irrigation and Drainage Paper No. 56, Rome, Italy, 1998.

[12] Gordon, N.D., McMahon, T.A. & Finlayson, B.L., *Stream Hydrology: An Introduction for Ecologists.* 2nd Ed, John Wiley and Sons, England, 2004.

[13] Kurc, S.A. & Small, E.E., Dynamics of evapotranspiration in semiarid grassland and shrubland ecosystems during the summer monsoon season, Central New Mexico. *Water Resources Research*, 40, Article ID: W09305, 2004.

[14] Davies, B. & Day, J., *Vanishing Waters.* University of Cape Town Press, South Africa, 1998.

[15] Schaffer, C.H., Investigating the Importance of Land Cover on Evapotranspiration. Published Msc Thesis in Geogaphy, Northern Illinois University, 85p, 2005.

[16] Mölders, N., *Land-Use and Land-Cover Changes: Impact on Climate and Air Quality.* Springer, 210pp, 2011.